I0427056

PRAWN FARMING PROSPERITY

Cultivate Delicious Prawns for Profitable Aquaculture

Dive Into The World Of Freshwater Prawn Farming And Boost Your Income With Expert Techniques

Dr. Fabian Felicity

CHAPTER ONE

Introduction

Prawn farming, often known as shrimp farming, has gained popularity as a profitable aquaculture business. This sector has grown rapidly as worldwide demand for prawns has increased.

Whether you're a seasoned farmer trying to diversify or a beginner interested in aquaculture, prawn farming may be a profitable undertaking.

In this detailed tutorial, we will go over the most important parts of prawn farming, from getting started to putting up your farming infrastructure.

Getting Started With Prawn Farming

Before beginning a prawn farming operation, it is critical to grasp the fundamentals. Prawns are crustaceans from the Penaeidae family, and effective aquaculture requires careful consideration of many criteria.

First and foremost, you must determine which prawn species you wish to produce. Different species have different needs, thus your decision will influence your farm's overall performance.

When starting started, look into the market demand for various prawn species. Pacific white shrimp

(Litopenaeus vannamei), tiger prawn (Penaeus monodon), and freshwater prawn (Macrobrachium rosenbergii) are among the most popular aquaculture species.

Each species has unique growth rates, environmental preferences, and commercial worth. To make an educated selection, consider water temperature, salinity, and the availability of resources in your selected site.

Understanding The Prawn Species

Prawn species have unique traits, and knowing their biology is critical for effective cultivation. Pacific white shrimp, for example, are noted for their resilience to a variety of environmental circumstances and rapid development. Tiger prawns, on the other hand, need special circumstances and are more prone to illness. Freshwater prawns flourish in freshwater surroundings, as their name implies, and are an excellent alternative for anyone exploring inland farming.

Consider the life cycle of your selected prawn species. Before reaching adulthood, prawns go through various developmental stages, including the larval and post-larval phases.

Successful prawn farming often requires providing ideal

circumstances for each growth stage. This might incorporate specialized nursery systems for larvae and grow-out ponds for maturity.

Choosing An Ideal Location For Prawn Farming

The success of your prawn farming operation is heavily reliant on finding the correct site. Prawns are very sensitive to their surroundings, and elements such as water quality, temperature, and salinity have a significant impact on their development.

Coastal locations are popular for prawn farming owing to the availability of salty water and accessibility to markets. However, inland places with adequate freshwater supplies might also be feasible for certain prawn species.

Water quality is essential in prawn farming. Regular testing for characteristics such as pH, dissolved oxygen, and ammonia levels is required to maintain a healthy environment for prawns. Furthermore, the water supply should be devoid of pollutants and toxins that may harm prawn development.

Investing in adequate water management measures, such as aeration and filtration systems, is critical to maintaining ideal conditions.

Climate is also an essential aspect. Prawns flourish in tropical and subtropical settings, and temperature variations may affect their development. Choose a place with a consistent environment that matches the temperature demands of your chosen prawn species.

Monitoring weather patterns and seasonal variations is critical for predicting any negative impacts on your prawn farm.

CHAPTER TWO
Setting Up Your Prawn Farming Infrastructure

Once you've chosen a good site, the following stage is to build the infrastructure for your prawn farm. Ponds, hatcheries, and processing facilities are key components. Pond size and design will be determined by your operation's scale and the prawn species you choose.

Pond preparation is a vital part of prawn farming. The process includes cleaning the ground, ensuring good drainage, and lining the pond to avoid water seepage. Ponds should be created to meet the unique requirements of your preferred prawn species, such as depth, aeration, and water exchange rates. Introduce natural substrates and flora to provide a suitable home for prawns while also encouraging the establishment of natural food resources.

Hatcheries are essential in prawn farming because they provide a regulated environment for the prawns' early growth. To guarantee healthy larval development, hatcheries must be managed properly, which includes regulating water quality, temperature, and nutrition. Consider purchasing specialized equipment, including larval raising tanks and feeding systems, to improve the efficiency of your hatchery.

Processing facilities are required for preparing prawns for sale. This comprises sorting, grading, and packing prawns according to market specifications. Adequate processing facilities assist in preserving the quality and freshness of prawns, increasing their market value. In addition, adequate waste management methods should be undertaken to reduce environmental impact.

To summarize, shrimp farming is a dynamic and potentially profitable activity that demands careful planning and consideration of several elements. From selecting the correct prawn species to determining the best site and establishing the required infrastructure, each step is critical to the success of your shrimp farming venture.

Understanding the unique features of various shrimp species, as well as investing in adequate infrastructure and management procedures, will help you establish a sustainable and lucrative prawn farming company.

Water Quality Management For Prawn Farming

Prawn farming, commonly known as aquaculture, has made a considerable contribution to world seafood output.

However, the success of prawn farming is inextricably linked to the control of water quality in the farm environment. Water quality has a significant impact on prawn health and development, thus aquaculturists must employ effective measures to maintain ideal conditions.

Water quality management includes controlling characteristics such as dissolved oxygen, pH, temperature, salinity, and nutrient concentrations. Dissolved oxygen is critical for prawns' respiratory health, and maintaining enough amounts ensures their survival. pH levels should be monitored and controlled regularly to avoid swings that may stress and harm prawn development.

Temperature control is an important part of water quality management. Prawns are ectothermic, which means that their body temperature is affected by the environment. Maintaining a consistent and appropriate temperature range improves metabolic functions and overall health.

Salinity levels must also be carefully monitored since prawns are sensitive to fluctuations in salt content. Proper nutrient management, including nitrogen and phosphorus control, aids in water pollution prevention and aquatic ecosystem health.

To accomplish successful water quality control, prawn producers use a mix of monitoring tools and best practices. Continuous water quality monitoring systems give real-time data on important parameters, enabling farmers to make timely modifications.

Furthermore, using sustainable farming techniques, such as efficient waste management and avoiding overfeeding, helps to maintain a balanced and healthy aquatic ecology for prawns.

Feeding And Nutrition Tips For Healthy Prawns

In aquaculture, feeding, and nutrition are critical to prawn growth and development. To achieve optimum development rates and general health, prawns' nutritional demands must be carefully monitored. Prawns are omnivorous organisms with varied food needs throughout their life cycle.

As a result, developing and supplying well-balanced and species-specific diets are critical components of successful prawn farming.

Commercially available prawn diets are designed to suit the nutritional needs of various life stages, including larvae and adults. These meals usually include a combination of protein, fats, carbs, vitamins, and minerals.

Protein is an essential component since it promotes muscle building and general growth. Essential fatty acids, a kind of lipid, help to meet energy needs and improve reproductive function.

Feeding tactics should take into account the individual needs of prawns at various growth phases. Larvae, for example, need high-nutrient diets to sustain fast development in the early stages. As prawns develop, feed composition may need to be adjusted to reflect variations in nutritional demands. Overfeeding should be avoided since it not only increases production costs but also has a detrimental influence on water quality due to excess fertilizer intake.

In addition to manufactured feeds, producers often use natural food sources in prawn diets. This may include algae, plankton, and other tiny creatures found in agricultural environments. Farmers may increase the nutritional variety of prawn diets while also promoting a more sustainable and cost-effective feeding technique by using the natural food web.

CHAPTER THREE
Disease Prevention And Health Management At Prawn Farms

Disease outbreaks represent a substantial hazard to prawn farming operations, possibly causing economic losses and environmental damage. Disease prevention and health management measures are critical to maintaining a profitable and resilient prawn farming sector.

Biosecurity measures are an important part of disease prevention. These precautions are intended to prevent the introduction and spread of diseases inside prawn farms. Strict quarantine methods for incoming animals, frequent health tests, and limiting access to agricultural facilities are all necessary activities. Farmers may improve the environment for prawns to flourish by reducing the danger of disease introduction.

Regular health monitoring is another important aspect of illness prevention. Observing shrimp behavior, growth rates, and general look may aid in detecting early symptoms of sickness.

Furthermore, regular sampling for laboratory examination enables the detection of possible diseases before they create widespread problems.

Proactive health management measures may include the use of vaccinations or probiotics to boost prawn immune systems and minimize vulnerability to illnesses.

Water quality management, as previously noted, is intimately related to illness prevention. Properly maintained water quality decreases stress on prawns, making them more resistant to illness. Adequate filtration systems, regular water exchanges, and the removal of organic material all contribute to a cleaner, healthier environment for prawns.

In the case of a disease epidemic, timely and efficient response measures are required. Isolating sick individuals, following treatment procedures, and, if required, consulting with aquatic veterinarians may all assist reduce the effect of illnesses on the prawn population.

Optimizing Growth And Reproduction In Prawns

Optimal growth rates and reproductive success are the major goals of prawn farming. Many variables influence prawn growth and reproductive success, and farmers use a variety of tactics to improve these features.

Maintaining appropriate water quality and providing well-balanced food are critical for optimal prawn development. Additionally, environmental parameters including as stocking density, habitat architecture, and water circulation are critical. Overcrowding may cause stress, competition for resources, and an increased risk of illness. Farmers must carefully regulate stocking densities to establish a favorable environment for development.

Reproductive success is critical to the viability of prawn farming businesses. Understanding the reproductive biology of prawns and ensuring favorable circumstances for mating and egg development are critical issues. Some prawn species have special mating needs, such as the availability of appropriate surfaces for egg attachment or the simulation of environmental signals to initiate reproductive activity.

Selective breeding is a prevalent approach in prawn farming to improve desired qualities including growth rate, illness resistance, and reproductive success. Farmers may establish better-adapted prawn populations by selecting breeding individuals with these qualities.

Harvesting Techniques For Maximum Yields

Harvesting is the last step of the prawn farming cycle and requires careful planning and execution to achieve optimal production and product quality. Prawn farming employs a variety of harvesting processes, each with its own set of pros and disadvantages.

Nets and traps are a frequent harvesting technique. These are carefully put in shrimp ponds, and when the prawns reach the required size, the nets are drawn to catch the mature ones. Careful scheduling is essential to prevent picking prawns that are too little or too big, resulting in ideal marketable sizes.

Another approach includes emptying ponds and collecting prawns as the water level drops. This strategy is especially successful in large farming systems because prawns have plenty of room to travel to the middle of the pond as the water drains. It is critical to properly control the water level to prevent straining the prawns throughout the harvesting procedure.

In more modern aquaculture systems, mechanical harvesters may be used. These devices employ suction or pumping mechanisms to effectively harvest prawns. While automated harvesters may drastically decrease manpower needs, their usage must be carefully planned to avoid harm to prawns and agricultural equipment.

Post-harvest management is an important factor that directly affects the quality of prawn products. Rapid cooling and good storage avoid product degradation and guarantee that prawns reach the market in peak condition. Some farmers also set up on-site processing facilities to clean, sort, and package prawns right after harvest, which adds value to the finished product.

To summarize, water quality control, feeding and nutrition, illness prevention, growth optimization, and harvesting procedures are all interwoven components that contribute to the success of prawn farming. Prawn farmers who use holistic and sustainable practices may not only increase production but also secure the long-term survival of their enterprises. The delicate balance of these elements emphasizes the significance of educated decision-making and ongoing monitoring in the dynamic area of prawn farming.

CHAPTER FOUR
Processing And Packaging
Your Prawn Harvest

Prawn farming is a profitable operation that needs careful consideration of several factors, with processing and packaging playing a critical part in deciding the success of your company. Efficient prawn harvesting guarantees the delivery of a high-quality product to the market, which in turn influences customer satisfaction and market competitiveness.

Processing Techniques

Effective processing processes are required to keep prawns fresh and of high quality. The first stage is cautious picking to reduce stress on the prawns. The utilization of specialist equipment, such as aerated tanks, provides a smooth transition from pond to processing facility. Proper handling at this stage makes a considerable difference in the quality of the finished product.

When prawns arrive at the processing plant, they are subjected to stringent quality control inspections. Sorting by size and quality is critical for consistency in the finished product. Following that, the prawns are thoroughly washed and deveined. Using sophisticated processing technology improves efficiency while also reducing the danger of contamination.

Cooking is another important part of prawn processing. Different markets prefer cooked or raw prawns, therefore processing processes must be tailored appropriately. Properly cooked prawns not only exceed safety requirements but also contribute to a longer shelf life.

Packaging Innovations

The packing process is an important but frequently ignored aspect of prawn aquaculture. Packaging not only protects the goods during transit and storage, but it also works as a marketing tool. Innovations in packaging materials and designs may have a big influence on shrimp product marketability.

Using vacuum-sealed packaging helps to preserve the freshness of prawns by reducing exposure to oxygen, thereby preventing spoilage. Additionally, it increases the product's shelf life, allowing for wider distribution and reducing potential waste.

Labels and branding on packaging help to communicate important information to customers. Mentioning the source of prawns, farming procedures, and any certifications received promotes openness and confidence. Sustainable packaging options, such as biodegradable materials, meet the growing consumer demand for environmentally friendly products.

Marketing Strategies For Prawn Products

Once your prawns have been processed and packaged, you must implement effective marketing strategies to ensure they reach their intended audience and generate revenue.

Understanding Your Target Market

Before launching a marketing campaign, it is critical to conduct extensive market research. Understanding your target audience's preferences and needs enables the development of tailored strategies. Geographic location, cultural preferences, and economic considerations are all important factors to consider.

Creating An Online Presence

In the digital age, having a strong online presence is not optional. Create a professional website to showcase your prawn products, including high-quality images and detailed product information. Use social media platforms to connect with potential customers, share updates, and solicit feedback. Online marketing can help you reach beyond your local market.

Collaborating With Retailers And Restaurants

Building partnerships with retailers and restaurants can enhance your prawn sales. Collaborate with local grocery stores, supermarkets, and seafood markets to ensure your products are readily available to consumers. Restaurants can provide an avenue for showcasing the versatility and quality of your prawns, creating a buzz that extends to the wider community.

CHAPTER FIVE
Financial Management And Profitability In Prawn Farming

Effective financial management is essential for the sustainability and growth of your prawn farming business. Careful planning, budgeting, and monitoring of financial activities are key to maximizing profitability.

Cost Analysis

Conduct a comprehensive cost analysis that includes all aspects of prawn farming, from initial setup costs to ongoing operational expenses. This should cover expenditures related to pond construction, stocking, feeding, processing, and packaging. Identifying all expenses provides for improved pricing strategies and profit margin estimates.

Diversification And Value Addition

Consider broadening your product offerings to appeal to a larger clientele. This might entail offering value-added prawn goods such as marinated prawns, prawn-based sauces, or ready-to-cook prawn recipes. Diversification not only boosts income streams but also enhances your brand's attractiveness.

Efficient Resource Utilization

Optimizing resource use is crucial for cost reduction. Implement efficient feeding procedures, check water quality to limit illness risks, and invest in energy-saving equipment. Regular maintenance of ponds and equipment eliminates unforeseen malfunctions and guarantees the life of assets.

Environmental Sustainability In Prawn Farming

As the worldwide emphasis on sustainable practices develops, implementing environmental sustainability measures into prawn farming is both responsible and economically profitable.

Water Management

Effective water management is vital for sustained prawn farming. Implementing water recirculation systems decreases the need for excessive water consumption and minimizes environmental effects. Monitoring water quality periodically helps maintain a healthy environment inside the ponds, boosting the well-being of prawns.

Responsible Feeding Practices

Adopting sustainable and properly obtained feed for prawns is critical. Investigate and invest in alternative protein sources to reduce dependency on wild-caught fish for feed production. This not only decreases environmental effects but also meets customer demand for ethically farmed seafood.

Biodiversity Conservation

Preserving biodiversity in and around prawn farms benefits the overall health of ecosystems. Implementing buffer zones, growing plants, and minimizing the use of toxic chemicals are all activities that help to conserve biodiversity. Participating in community environmental projects helps to improve the image of your prawn farming company.

Conclusion

Successfully negotiating the challenges of prawn farming requires a comprehensive strategy that includes processing, packaging, marketing, financial management, and environmental sustainability. By focusing on these factors, prawn farmers may not only create high-quality goods but also build a robust and lucrative company in the competitive seafood market.

The combination of efficient processing, creative marketing, good financial practices, and environmental care lays the groundwork for long-term success in prawn farming.

www.ingramcontent.com/pod-product-compliance
Lightning Source LLC
Chambersburg PA
CBHW071015290526
45795CB00005B/1812